Farm Tools and Techniques

A Pictorial History

Farm Tools
and Techniques

A Pictorial History

Jonathan Brown

B. T. Batsford Ltd · London

First published 1993

© Jonathan Brown 1993

Typeset by Latimer Trend & Company Ltd, Plymouth, Devon

and printed in Great Britain by
The Bath Press, Bath

Published by
B. T. Batsford Ltd
4 Fitzhardinge Street
London W1H 0AH

A catalogue record for this book is available from the British Library

ISBN 0 7134 7499 8

Contents

Preface

In another book published by Batsford in 1989 (*Farm Machinery 1750–1945*) the development of the implements and machines used in farming was described. While mechanisation was proceeding, however, farm techniques reliant on human muscles and hand tools retained an important role, though declining as more processes were mechanised. This volume describes some of those handicraft techniques of the farm, and their tools, many of which are now to be found in the collections of museums up and down the country. In addition, attention is given to some of the rural manufacturing and servicing crafts that supplied and repaired the tools and equipment used on the farm.

The illustrations are all from the collections of the Rural History Centre, University of Reading, with additional acknowledgement to Dorset County Museum (71, 79), *Farmer & Stockbreeder* (8, 13, 41, 42, 88) and Hampshire County Museum Service (32, 92). Mrs Barbara Holden of the Rural History Centre and the members of the University of Reading's photographic service gave their usual unfailing attention in the supply of all the photographs.

1

Introduction

It is a common irony that changes in technology and the spread of mechanisation, initially at least, often lead not so much to a saving of labour, but to its redistribution. Mechanisation in one area increases productivity to such an extent that extra effort has to be put in elsewhere to create conditions suitable for these machines. This was the experience of agriculture in the nineteenth century. New implements such as drills, horse hoes, cultivators and improved ploughs raised standards of cultivation and made them more intensive. Yet many of the more traditional techniques and crafts reliant on hand tools, far from being discarded, were adapted to take their place alongside the machines. It was only with the complete mechanisation of farming after 1945 that these techniques were superseded.

Many operations of farming remained almost untouched by mechanisation until the late nineteenth and early twentieth centuries. Most prominent was harvesting, where the corn continued to be cut by hand and the crop gathered by hand rakes, then stooked, carted and stacked manually. Where the new implements were being used, fields were prepared so that they could work efficiently – drains were laid, hedges straightened, trimmed and maintained, all operations that continued to be labour intensive.

These techniques did not stand still. Harvesting methods changed during the course of the nineteenth century as scythes and bagging hooks frequently replaced the sickle and reap hook. New types of seed-broadcasting equipment were introduced. Techniques such as paring and burning had their times of popularity before declining into more residual operations. The tools themselves changed, under the influence of industrialisation. Tools that had always been bought from the local blacksmith or had even been home-made were, from the second half of the nineteenth century, increasingly supplied by larger manufacturers. Ransomes of Ipswich, known mainly for their ploughs, threshing machines and steam engines, saw the potential market for standardised hand tools, and in the 1850s were offering a range of spades and forks for drainage, manuring and other farm tasks. Ransomes soon dropped out of this business, which became dominated principally by the firms of toolmakers in Sheffield. These offered extensive ranges of agricultural forks, billhooks, turnip knives, sickles, scythes. They did their best to marry standardisation with regional tastes and particularities. Various 'local' patterns of knife and hook were being made into the 1930s and 1940s, although some were likely to have changed subtly over the years.

The growing importance of the Sheffield firms as suppliers of farm tools was a part of the process whereby farming was separated from the interdependent rural trading pattern that had sustained it. Farmers had been reliant upon the crafts of the village and market town. Most prominent were the blacksmith, the wheelwright and the carpenter. These made and mended the tools of the farm and, most important, maintained the horse power of the farm – shoeing horses, repairing wagons. Besides these, there were numerous other crafts and trades linked to the farming economy – the hurdle maker, the rake maker and the basket maker, who made the feeding baskets for cattle, for example. Farming thus was never an independent entity, but during the nineteenth and early twentieth centuries its dependence shifted from its own locality and region to a dependence on a wider industrial economy. With that shift many of farming's old tools, techniques and crafts declined or disappeared altogether.

1 *Old meets new at the haystack. A stream traction engine provides power for the mechanical elevator, while the team of men and women load the hay with their pitchforks. The photograph was taken early in the twentieth century.*

2

Cultivating the Land

In his 1860 text book, *British Agriculture*, Professor Donaldson wrote: 'Forks, spades and digging tools are used by the hand in digging ground, and in moving farmyard dung. The articles are indispensable on any farm, and form the first requisites of use.' The importance of hand tools and techniques was still being stressed at the end of the nineteenth century. W. Fream in the 1983 edition of his *Elements of Agriculture* noted that 'hand tools are as necessary as the larger implements on the farm'. Only in the twentieth century were hand tools deemed barely worthy even of a passing reference, although many had regular use in farming into the 1940s and 1950s.

Paring and Burning

Paring and burning was a particular operation in the preparation of land for cultivation whereby a thin layer, up to about three inches deep, was pared off the top of the soil. The spoil was gathered up and burned, so destroying the coarse vegetative matter. The ashes were spread across the surface of the field, where they either remained or were lightly ploughed or harrowed in. This was a practice known since at least the fourteenth century and was a feature of farming in Devon, as a result of which paring and burning was in other parts of the country often called devonshiring or denshiring. In the late eighteenth and early nineteenth centuries, when large areas of waste and common were being taken into cultivation, it became far more widespread, and was found in most parts of England and Wales. The benefits to be gained from paring and burning were much debated, with regular discussion in agricultural books into the 1860s and 1870s. Its advocates maintained that it was the most effective way of cleaning land, and the ashes gave

valuable manure for the succeeding crop. Arthur Young, a convinced supporter, thought paring and burning 'properly managed, is the most admirable of all improvements, and improperly, the most mischievous'. He cited the testimony of many farmers in support of the practice, such as the man in Lincolnshire who was 'decidedly of opinion, that it is the most expeditious way of bringing any land that has long been under rabbits, or any spontaneous growth, into cultivation'.

There were two principal methods of paring off the surface. One was to use a beating axe, a large adze, to chip and hack the sods out. This, in Devon, was known as hand beating. The other method properly skimmed or pared the surface using the hand tool most widely associated with paring and burning – the breast plough. This had a flat, pointed blade and, on most types, an upturned edge which performed as the coulter did on a plough. There was a long shaft and wide cross handle. The implement was pushed, in a jerky movement, until about a yard of turf had been undercut and could be turned over. It was hard work, 'the most slavish work of husbandry' thought William Marshall. Progress was slow – about a quarter of an acre a day was generally reckoned to be the limit for one man.

The breast plough reached its most widespread use during the Napoleonic Wars when extensive amounts of waste were being taken into cultivation. Thereafter paring and burning began to decline in popularity, but there were some areas where its practice continued as part of the regular course of arable farming, in breaking temporary pastures and even wheat stubbles. Parts of south-western England, the Cotswolds, Kent and the Fens were where paring and burning and, with that, the use of the breast plough, persisted throughout the nineteenth century.

Land Drainage

Well-drained land is vital to arable farming. There were two basic methods of field drainage: surface drainage, effected most commonly by ploughing up ridges that allowed water to run off into furrows on either side, and underdrainage. It was the latter that gained in importance, especially after the 1840s when cheap, mass-produced drainage tiles and pipes became available.

'No improvement can have greater or more immediate effects than this of draining; none that pays the farmer with more certainty,' wrote Arthur Young at the beginning of the nineteenth century when the practice was steadily becoming more widespread. The most usual means of underdraining at that time was to fill the bottom of a trench with stones or vegetable matter, such as brushwood, straw or stubble, to provide a channel through which water could run. These materials were still in use in the mid-nineteenth century, although by then tiles and pipes were becoming usual. Earthenware tiles and semi-circular pipes began to be used at the end of the nineteenth century. The invention in the 1840s of machines for making drainage pipes coincided with an upsurge of interest in the virtues of underdrainage. Landowners and farmers began to invest heavily in tile drainage, with many estates setting up their own tile-making operations. The government made loans available to encourage landowners to invest.

While the pipes were made by machine, much of the work of digging the trenches and laying the tiles remained a manual task. There had been drainage or trenching ploughs in use since at least the seventeenth century. Some could cut a trench about twelve inches deep. They tended to be heavy and cumbersome, however, and to require a great deal of horse power to draw them. Several types of mole plough were devised during the nineteenth century, bringing up to date an old principle of boring a tunnel through the soil. Steam power was applied to this technique, notably by John Fowler, who produced versions to be worked by portable steam engines and to be operated with his steam ploughing engines. He also produced a tile-laying machine that could be powered by a portable engine. Trench-digging machines of various types were devised by a number of inventors and exhibited at the shows of the Royal Agricultural Society of England.

The mole plough was the most successful of these implements, but none gained widespread use. It was not until the mechanical excavators of the twentieth century arrived that draining work became mechanised. Until then most drains were dug out manually using special spades with narrow blades suitable for cutting out the v-shaped wedge that was needed. The first furrow was often opened out with a plough, then the men set to work with the top spade. To complete the wedge they turned to the bottom spade, which had a narrower blade. The bottom of the trench was made smooth and clean using a long-handled scoop. A pull scoop was usually preferred for drawing out the mud from a wet trench. In drier conditions a push scoop might be used. The pipes could then be laid: another long-handled tool with a long straight hooked end could be used to drop the pipes in without the men having to get down into the trench.

Laying drains was a job mainly for the winter, in fields kept back for spring cultivation. There were farmers who preferred summer for this work, on the grounds that in the long days more could be accomplished, while carts and wagons could get across the land more easily to deliver pipes. There was general agreement, though, that spring and autumn were not good times for this work.

Sowing Seed

By the middle of the nineteenth century the horse-drawn seed drill was gaining widespread acceptance as an efficient means of sowing seed in rows. Broadcasting and dibbling both continued into the early years of the twentieth century. Dibbling, the alternative means of setting seed in rows, was popular in East Anglia. For sowing corn it was a novelty of the late eighteenth century. William Marshall reported that in Norfolk dibbling peas had been practised since 'time immemorial', but as a method of sowing wheat it had been introduced in about 1770 in south Norfolk. From there the practice spread across Norfolk, Suffolk and into Essex. Light soils were found to be most suitable. The virtues of dibbling were much debated. Wilson's *Rural Cyclopedia* of 1851 observed that 'it continues to provoke variety and conflict of opinion . . . Many persons insist that dibbled crops on even perfectly suitable lands are not perceptibly superior to drilled crops; and others not only assert their growth to be more vigorous . . . and their seeds to be plumper and heavier, but estimate them to have a superior produce in measured grain varying from two to twelve bushels per acre'.

Many farmers continued to have their seed dibbled throughout the nineteenth century. Some dibbling machines were devised, but on the whole the farmers preferred the manual method. It was, for the labourers most proficient at it, a valuable skill, for they were able to hire themselves at piece rates to all the local farmers and smallholders. Nine shillings an acre, plus beer, was a common rate in the late nineteenth century, and the quickest labourers could do half an acre a day.

By the third quarter of the nineteenth century broadcast sowing of corn by hand was becoming unusual except on small farms. For sowing grass, however, broadcasting remained common well into the twentieth century. Rhythm counted for everything when sowing by hand. It was usual for sowing to be done on the move. Then it became important for the hand to 'keep tune with your foot', as one old labourer commented in the 1930s. 'If the hand and one foot alternately do not move simultaneously the ground will not be equally covered, and a strip left between the casts,' wrote Henry Stephens.

In Stephens' native Scotland the most usual container for the sower to carry the seed was a sowing sheet, worn in a fashion similar to a sling and wound tightly round the left arm in which it was cradled. The English preferred a wicker sowing basket. 'Such an instrument,' remarked Stephens, 'no doubt, answers the purpose of the sower, or it would not have been so long in use; but for my part, I much prefer the comfortable feel of the linen sheet to the hard friction of the wicker basket.'

With the seed in the linen or the sowing basket the sower would broadcast with one hand, casting the seed to the one side (usually the right) over half a ridge. The other half ridge would be covered when the sower came back across the field. There were many workers, however, who preferred to use both hands, making casts alternately with right and left hand, to sow a whole ridge on one passage of the field. Henry Stephens did not like this idea: 'I can see no advantage attending this mode of sowing over the other; but, on the contrary, a considerable risk of scattering the seed unequally.' For sowing two-handed the seed container needed to be strapped so that both hands could reach it freely. The most suitable type was the large box, often called a seedlip. It was made usually of deal (later versions were of galvanised iron) often shaped so as to fit snugly round the sower's middle.

The sowing fiddle was a new tool of American origin introduced into Britain in the late nineteenth century. The bow, which gave it its name, was drawn with each stride and this set a wheel revolving which scattered seed forward of the sower. The seed could be projected over quite a wide arc – up to 30 feet – and the experienced sower could expect to cover about four acres an hour.

Weeding

Weeding and singling crops by hand remained common into the twentieth century. 'Though the same work could be done with a horse-hoe or even a harrow at a much cheaper rate, undoubtedly the hand work gives better results,' wrote Primrose McConnell in 1911. The hoe was the principal tool, of which there was a great variety of patterns. William Marshall in the 1790s observed women in Devon going out

to hoe turnips with hoes fashioned from old scythe blades. Such locally-made tools remained common for many decades, but by the end of the nineteenth century the large manufacturers of edge tools based in Sheffield dominated the market for new implements, each offering several different types of hoe.

Preference varied, partly from region to region, but also according to the type of work being done. For hoeing of corn crops Primrose McConnell advised a blade set at an acute angle to the shaft, the better to stir the soil to about an inch deep. For singling root crops, on the other hand, he said the blade should be set square to the handle to give the precision of work required. The blade attached to the handle with a curved, 'swan' neck was much favoured – Fream thought it the best, and McConnell approved. The size of blade varied, too, according to crop; four-and-a-half to five-and-a-half inches wide for weeding wheat and barley, six to eight inches for peas, and nine inches for beans and root crops were recommended by Fream.

Harvesting

By the end of the nineteenth century harvesting was becoming steadily more mechanised. The reaping machine had found favour for cutting corn from the 1860s onwards. It was succeeded by the binder, which tied the sheaves as well as cut the corn. It began to come into general use during the 1890s. By 1900 about eighty per cent of the British corn harvest was being cut by machine. The hay crop likewise was increasingly being cut by mowing machines.

There remained at the beginning of the twentieth century a fair amount of harvesting by hand. Even on farms predominantly mechanised, the scythe and the hook might be used for opening out fields to provide access for the binder and for cutting badly laid crops.

There were three methods of cutting the corn by hand. Firstly reaping, for which the tools were the sickle and the reaping hook. The two were similar – the sickle had a serrated blade, the reaphook a smooth-edged one. Secondly, bagging – cutting with the heavy, smooth-bladed bagging hook. Thirdly, mowing with the scythe.

In 1813 R. W. Dickson, discussing the relative merits of these methods of cutting the grain crop, advised the readers of *The Farmer's Companion* that where labour was scarce, the best course would probably be to use the scythe. If labour was plentiful then the sickle was likely to be the best tool, for with it the crop would be 'placed with more exactness and regularity, and of course is capable of being bound up with greater ease and facility. And it is probable that there is less loss of grain incurred in the operation'.

At the time Dickson was writing, wheat and rye were cut mainly with the sickle or reaping hook. The scythe, he observed, was used in some early districts, such as the Midlands. The scythe was used more frequently for cutting barley and oats in southern England; hay was almost invariably mown. By 1870 there had been quite a change in the wheat harvest for, on the eve of mechanisation, two-thirds or more of the crop was now being cut by scythe. The availability of labour, to which Dickson referred, had largely determined this change. Workers were leaving the villages, while more wheat was being grown. There was ever more anxious anticipation amongst the farmers as harvest approached and they waited to see how many of the itinerant Irish harvesters would arrive. With labour more scarce farmers would turn to the scythe and bagging hook as tools. The skilled scythesman could cut more than an acre of wheat in a day compared with the rate of about a third of an acre that the reaper could manage. The bagging hook came between these two, at about an acre a day per man.

There were considerable differences of technique in cutting with the different hand tools. Reaping was a slow and careful process. The reaper stooped down to grasp a handful of corn, which he cut slowly. With a sickle he would use a gentle sawing action. The reap hook, with its smooth blade, required more of a chopping action, but, again, it was slow and deliberate. Each handful, once cut, was placed upon a band of straw rope laid out ready on the ground. When enough handfuls had been gathered in this way they were tied into a sheaf, either by the reaper himself or, more likely, by somebody else following behind. The reapers might be organised into regular groups, each with its supervisor. This was especially common in the north of England and Scotland. The Scots called the team a bandwin (of seven) working three to a ridge with the seventh man the 'bandster' who supervised and gathered sheaves up into stooks. Such systems were not so regularly followed in other regions.

The reaper might cut the corn close to the ground or up to six inches up the stalk. The balance of opinion was generally in favour of cutting close to the ground, for the longer stalks were reckoned to be easier for threshing, and there was more straw, which could be used as bedding for livestock. There was also the long stubble to dispose of, either by burning or by mowing it off. Cutting low with the sickle had the disadvantage of being much slower and, with more stooping, was doubtless more severe on the backs of the labourers.

The scythe and the bagging hook were perhaps not tools for the purist, especially not the bagging hook – its strong, slashing action and the untidy sheaves it left being regarded by many as crude compared with the gentle sickle.

But there was no denying the advantage of speed that this and the scythe both possessed. Both these tools, too, cut the corn close to the ground and with less back-breaking effort.

Stooking, Carting and Stacking

Until the combine harvester came along the cut corn had to be gathered, bound into sheaves, stooked in the field to dry and then carted off to the stackyard. These were all manual tasks, and the only one to be eliminated in the earlier stages of mechanisation was the tying of the sheaves, which the self-binding reaper did. The machine used wire, later twine, to bind the sheaves. The men used straw bands – two or more wisps of straw simply twisted together to form a bond strong enough to hold the sheaf. The bands were often prepared in advance of the day's cutting. Wilson's *Rural Cyclopedia* of 1847 advised that 'The bands ought to be laid in the morning, that they may not crack; for, after the sun is up, the strain loses its elasticity, and cannot properly be twisted, but becomes brittle and liable to break below the ears'.

For tying small sheaves a single band was enough – one length of straw around the sheaf and knotted head to butt. Large sheaves needed a double band: two lengths of straw were joined at the butt end to form the longer band, and the main knot was then head to head.

The sheaves were stooked for up to twenty days. Six, eight or ten were the usual numbers of sheaves to a stook, arranged in two rows meeting at the heads and slanting outwards to the butt ends. This allowed a free passage of air for drying. In the wetter regions other methods of stooking were often employed in order to give extra protection against the rain.

Once the corn was dry, it was carried off the field to be stacked. This had not always been the practice. In medieval times the crop had been stored in a barn, but by the eighteenth century higher yields and the greater quantities of straw meant that the capacity of the barns had been far outstripped and stacks outdoors were necessary. On the large farms these made an impressive sight, a dozen or more stacks together in the stackyard.

Shapes and sizes of stacks varied. There were round stacks, oblong stacks (some long, some short) and stacks that were almost square. There were differences in height to the gables as well as in the shaping and rounding of the ends. All of these varied from place to place. In general the northern preference was for round stacks, while oblong ones predominated elsewhere, but these divisions were not hard and fast.

Whatever its shape, the stack had to be constructed so as to provide the maximum protection against rats, mice, birds and the weather. A stack built directly upon the ground was given protection against rising damp by a layer of loose straw. An extra tier of large stones might then be added, on which timber battens could be placed to form a frame, thus lifting the stack a bit further off the ground. Staddle stones, a more effective way of raising the stack from the ground, were introduced during the eighteenth century. Their broad-spreading mushroom-shaped caps provided added protection against mice, who were prevented from climbing up into the stack. The final sophistication was the cast-iron rick stand, which began to find favour from the 1840s and 1850s. These were often recommended as enabling crops to be secured the better in wet seasons because of the free circulation of air underneath the stack.

To keep birds out stacks were built with the butt ends of the sheaves forming the outer wall. The most determined sparrows, however, would not be deterred: they would keep tugging at a stalk until eventually they could get at the ears. The outer walls of some stacks were trimmed off, giving emphasis to the fact that rick-building was something of a fine craft. Professor Donaldson described the operation in 1860: 'sheaves are singly delivered to the rick from the hand fork of the carter or waggon driver, and a boy throws them to the rick builder who places each sheaf in the proper position, and presses the bulk firmly with his knees, which are protected by a piece of strong leather or sheep skin.' Successive layers of sheaves were thus carefully placed upon each other, each layer being built with a slope up to the middle so that the centre of the stack was always considerably higher than the walls. This helped in shaping the top of the stack up to its apex. It also anticipated the natural settling of the stack: a stack built dead level was likely to settle and the middle sink. The result would be that rain would go straight into the centre instead of running off harmlessly down the outside.

William Fream, in his well-known text book, *The Elements of Agriculture*, commented that 'a well-built stack suffers little from rain even when unthatched, while one which is badly constructed suffers considerably when it is thatched'. It was, in fact, the usual practice to thatch stacks, both of hay and corn, as a protection against the weather. In addition, a good covering of thatch gave a smart finishing touch, often enhanced with decorative finials or other details.

Summer and early autumn were thus particularly busy times for thatchers. In Suffolk, for example, thatchers could expect to work on as many as three hundred hay and corn stacks in a year. With all the farmers wanting their ricks to be thatched at the same time, all who did any work at thatching were in demand. There were all the independent craftsmen, including many part-timers, men whose skills at

thatching were employed mainly on rick-thatching and who for the rest of the year had other business. Because thatching a stack was not such a skilled operation as thatching a house, there were in addition many farm labourers, especially on large farms, who were able to take on this work.

Thatching a stack was quite straightforward and, with two men at work together, a stack could be finished in a day. Straw was prepared in advance by being soaked to soften it, thus to make it 'lie'. Handfuls of this straw were spread neatly over the top, smoothed with a comb, and secured with spars and hay bonds.

The bonds or rope of hay were made by the thatchers themselves using a rope twister or throw crook. With a few lengths of straw fastened on to the hook of the twister, the handle was cranked to twist them into a bond. More lengths were added as the spinner walked backwards turning the twister all the time. The thatching spars were made of hazel by one of the woodland craftsmen, often as a sideline by the maker of wattle hurdles. Spars were made by taking a straight length of hazel about three feet long and cleaving it with a billhook. Each half-round was cleft again so that there were four strips out of the original rod. The strips were each bent double to form a large broach (spars were called broaches in Norfolk).

In the late nineteenth century the rick thatcher's art met its first serious challenge – from galvanised iron stack covers. Rider Haggard tried them one year when he was short of straw for thatching, and found that 'although not ornamental [they] are exceedingly useful . . . One disadvantage of these roofs is that the stacks must be built to fit them, and another that they undoubtedly look ugly in a farmyard, although this fault might be mitigated by painting them straw colour'. Besides the iron ones, farmers began to look to other cheaper alternatives to thatching, such as tarpaulin sheeting. What really brought about the demise of rick thatching was the combine harvester and straw baler.

Threshing

Writing in 1899, Rider Haggard could remember 'seeing the flail used from time to time, the last occasion being not more than fifteen years ago. From a flail to a modern steam-thresher is a long stride, and the time and labour saved by the latter are almost incalculable'. By this time mechanisation of threshing was all but complete. About eighty per cent of the corn harvest was already being threshed by machine by 1870. Threshing with the flail was largely confined to small farms, and a few particular needs: there were farmers in East Anglia who preferred to thresh barley for malting by hand on the grounds that fewer ears were likely to be damaged that way. In these ways the flail lingered in use such that, in the 1930s, C. Henry Warren could note that there were still rare occasions when one might be 'fortunate enough to catch the sound of the rhythmic beat of the flail where some old farmer is threshing out his beans'.

Threshing by hand was a slow process that was carried on throughout the winter months. A day's output could be no more than seven bushels of wheat. It was monotonous, tiring and wearisome work, but it did have the virtue of, for the most part, being work under the cover of the barn roof. Most barns had a specially-prepared threshing floor immediately behind the large double-doors through which the wagons brought the corn in from the stack.

The flail was a tool with a short life. Not even the most durable of woods could survive more than a few years of this continual beating. Crab, blackthorn and holly were among the resistant woods favoured for making the swingle, the shorter of the two sticks and the one that did the beating. The handstaff, the longer and thinner stick, up to four or five feet long, was of ash or some similar lighter wood. The handstaff was less likely to wear out, and was likely to have a succession of new swingles attached to it by the leather thong. The labourers themselves tended to make their own flails.

After the threshing came the finishing processes of winnowing and dressing. Straw, stones and corn that had escaped the flail unthreshed could be separated by a simple riddle. Separating the chaff from the grain had traditionally relied upon the natural strength of the wind to blow the light chaff away from the heavy grain. A shovelful of threshed corn thrown into the air would result in the chaff being blown to the side while the grain fell straight back to the ground. Indeed, the wind could automatically sort the grain for the best, the heaviest sample would travel the furthest, the medium sample would land a bit closer, and the 'tail' corn fit only for feed for animals would hardly travel at all. A large, shallow winnowing basket was the alternative to the shovel for tossing the corn up.

Artificial means of generating a wind for winnowing by large fans had been devised by the ancient Chinese and knowledge of these had been brought to Europe in the sixteenth and seventeenth centuries. Large canvas sheets on a revolving frame cranked by hand was a rather cumbersome method. Far more efficient was the fan of wooden blades enclosed in a box. Winnowing machines worked on this principle, powered by hand, horse gear or steam engine, became common during the late eighteenth and early nineteenth centuries.

Dressing the corn was done by riddling out the refuse, and various types of small, hand-powered machines using flat-bed or rotary screens were taken up during the nineteenth century. Many farmers continued to use them for additional cleaning of grain before market, even after the advent of the large combined threshing machine from 1848. These machines incorporated the winnowing and dressing with threshing in one combined operation, thus changing for ever the work of the threshing barn.

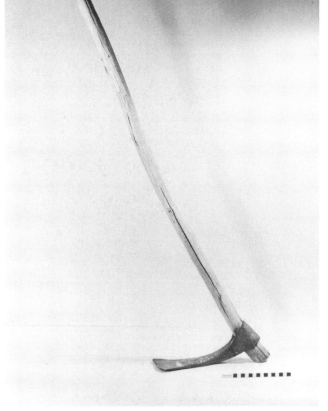

2 *'Spade work is expensive, but it is effective amongst roots and stones, and the ground is put by it in so prepared a state the succeeding operations are executed with ease and satisfaction.' These words of Henry Stephens from the mid-nineteenth century could apply nearly a century later. Here land is being cleared in Oxfordshire during the Second World War using mattocks and twibills (double-bladed axes).*

3 *A mattock, a large adze used primarily for clearing land. The beat axe, referred to by William Marshall in connection with paring and burning in Devonshire, was a similar tool.*

4 *A breast plough. The pressure to force it through the ground came from the man's thighs. As can be seen here, it was common to wear wooden pads, held together with leather straps, to protect the thighs and upper legs.*

5 *Heavy going: digging out a water furrow for surface drainage in a field of stiff clay.*

6 *A clay spade, used mainly for cutting surface drains. These were often known as willow spades, as that was the wood from which they were usually made. These wooden spades were often preferred to metal ones because heavy clay soils did not cling so much to them. They were made from a single piece of wood – leaving no joints for water to seep in and destroy the spade – and then sent to the blacksmith who shod the edge of the blade with iron strips. Some of the larger manufacturers of tools also produced these spades, offering them in their catalogues at least until the First World War.*

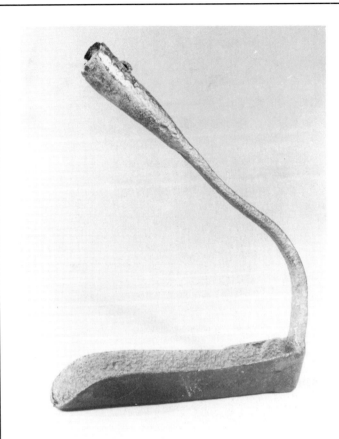

7 *A draw scoop for cleaning drainage trenches. A wooden handle about five feet long was attached to the iron scoop.*

8 *Clay pipes being laid out alongside a drainage trench as the bottom is being dug out with the narrow spade. The scene is in Northumberland during the Second World War.*

9 *Shadrach Best of Shottery, near Stratford upon Avon, Warwickshire, demonstrates the art of dibbling. With an iron dibble in each hand the worker walked backwards making a hole with each dibble at intervals of three to four inches. A quick twist of the wrist as the hole was being dug ensured that the dibble would come out clean and free from mud. A piece of rag on a stick on the headland was the sight marker for the rows. The dibbled holes were filled with seed by others following behind, often children, who earned up to 6d a day at this.*

10 *Sowing seed broadcast from a wicker basket, the harrows following behind. Westmorland, 1940.*

11 *Broadcasting survived as a means of distributing fertiliser more generally than it did for sowing seed. Fertiliser is being distributed here from a galvanised seedlip. The man is using both hands, so the container is held firmly in front of his body by straps.*

12 *The fiddle was for seeds only rather than for fertiliser, and is being used here for undersowing grass seed. The sack held about half a bushel of seed. It was important to hold the fiddle level in order to get good distribution.*

13 *Grassland improvement in the Denbighshire hills in 1942 brings out two methods of broadcasting. The fiddle is being used to distribute the seeds and the men with the seedlips are adding a dressing of the fertiliser, basic slag.*

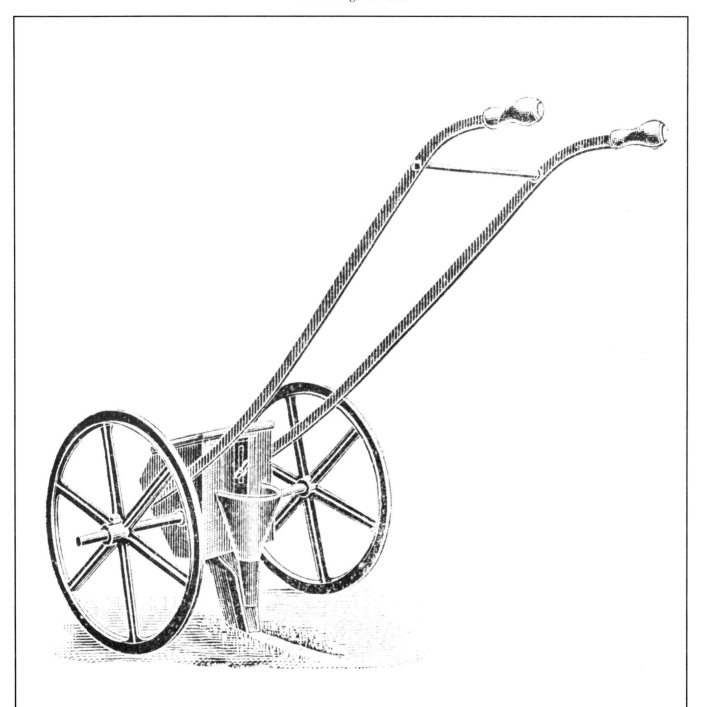

14 *Small hand-powered seed drills offered small-scale farmers and market gardeners an alternative to broadcasting or dibbling without the need for a large horse-drawn drill. Most of these implements sowed the one row only, and there were many versions intended for different types of seed. This example is from a late-nineteenth century catalogue by Reeves of Bratton, Wiltshire.*

15 *The work of gathering manure from the stalls and yards, and spreading it on the fields required several types of fork, shovel and drag. A manure drag is being used here to unload the cart in the field.*

16 *Heads down: the hoeing gang singling sugar beet in 1944.*

17 *Often it was necessary to get right down on hands and knees, using short-handled hoes and weeders, for crops such as onions, which are being singled in this field.*

18   *While the industrial makers of tools made inroads into the market, there remained many farmers and farm workers who preferred the home-made and blacksmith-made tools fashioned to meet their particular requirements. This swan-neck hoe is an example from Oxfordshire.*

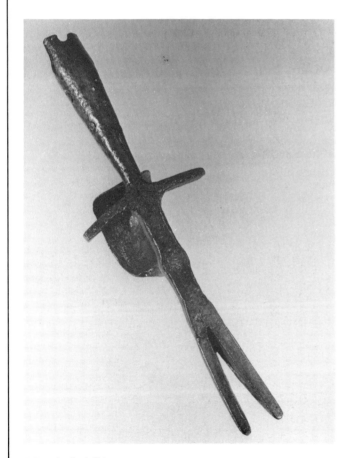

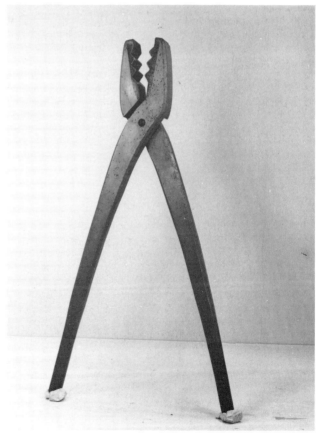

19   *A dock lifter.*

20   *These wooden weeding tongs are about two feet nine inches long. They have a coat of red paint, and the teeth make them look as though they really mean business.*

21 (above) *A busy scene in the hay field early in the twentieth century, the swaths being raked up and pitched on to the wagons.*

22 (below) *Turning the swaths of hay with rakes.*

23 *Manual hay drags such as this continued in use long after the introduction of horse-drawn rakes in the mid-nineteenth century. With heads three or four feet broad and long tines, drags were used in gathering the hay for loading the wagons or building the stack in the field. They were often made by the rake maker entirely of wood, while other types had metal tines.*

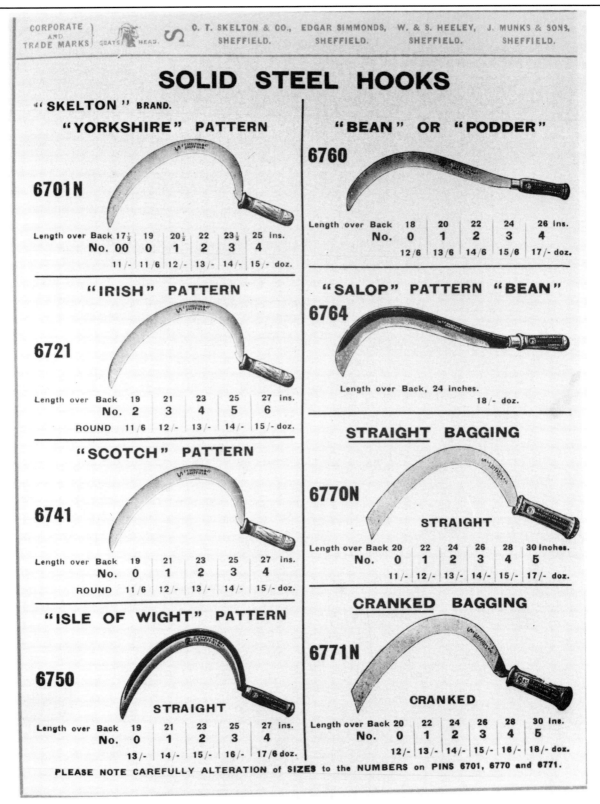

24 and 25 *The range of sickles and hooks for harvesting, including those intended for cutting beans, peas and furze, offered by C. T. Skelton & Co. in 1914. Cranked handles were intended to enable the harvester to keep his low cutting strokes more parallel to the ground.*

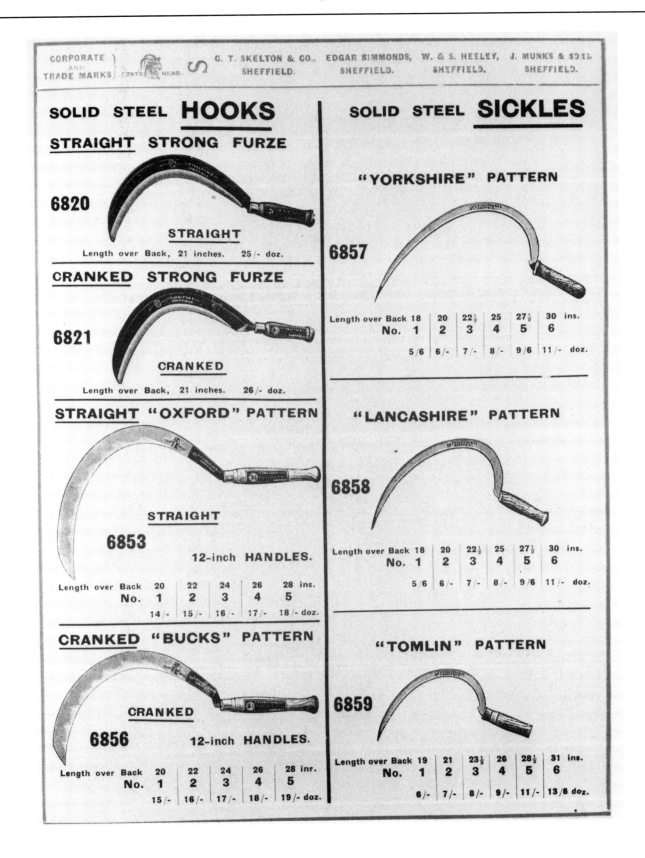

CORPORATE AND TRADE MARKS

C. T. SKELTON & CO., SHEFFIELD.   EDGAR SIMMONDS, SHEFFIELD.   W. & S. HEELEY, SHEFFIELD.   J. MUNKS & SONS, SHEFFIELD.

SOLID STEEL HOOKS

STRAIGHT STRONG FURZE

6820

STRAIGHT

Length over Back, 21 inches.   25/- doz.

CRANKED STRONG FURZE

6821

CRANKED

Length over Back, 21 inches.   26/- doz.

STRAIGHT "OXFORD" PATTERN

STRAIGHT

6853   12-inch HANDLES.

Length over Back	20	22	24	26	28 ins.
No.	1	2	3	4	5
	14/-	15/-	16/-	17/-	18/- doz.

CRANKED "BUCKS" PATTERN

CRANKED

6856   12-inch HANDLES.

Length over Back	20	22	24	26	28 inr.
No.	1	2	3	4	5
	15/-	16/-	17/-	18/-	19/- doz.

SOLID STEEL SICKLES

"YORKSHIRE" PATTERN

6857

Length over Back	18	20	22½	25	27½	30 ins.
No.	1	2	3	4	5	6
	5/6	6/-	7/-	8/-	9/6	11/- doz.

"LANCASHIRE" PATTERN

6858

Length over Back	18	20	22½	25	27½	30 ins.
No.	1	2	3	4	5	6
	5/6	6/-	7/-	8/-	9/6	11/- doz.

"TOMLIN" PATTERN

6859

Length over Back	19	21	23½	26	28½	31 ins.
No.	1	2	3	4	5	6
	6/-	7/-	8/-	9/-	11/-	13/6 doz.

26 *The harvest field, from the* Illustrated London News,
*1858. Women regularly were employed cutting the corn
with the sickle. The heavier bagging hook and scythe were
invariably men's tools, the women now doing the work of
gathering and tying the corn into sheaves only.*

27 *A harvester with a bagging hook, a large, heavy hook
with a broad blade, weighing up to about four pounds. The
crooked stick in his left hand was used to draw some of the
corn forward so that he could cut smoothly close to the
ground. The bagging hook was used in southern and
south-midland counties of England and the Welsh border
districts (usually only for wheat). It was unknown in
northern districts.*

28 *Using a bagging hook to deal with a badly laid crop in the 1930s.*

29 *A scythe with the English-pattern s-shaped shaft (or sned).*

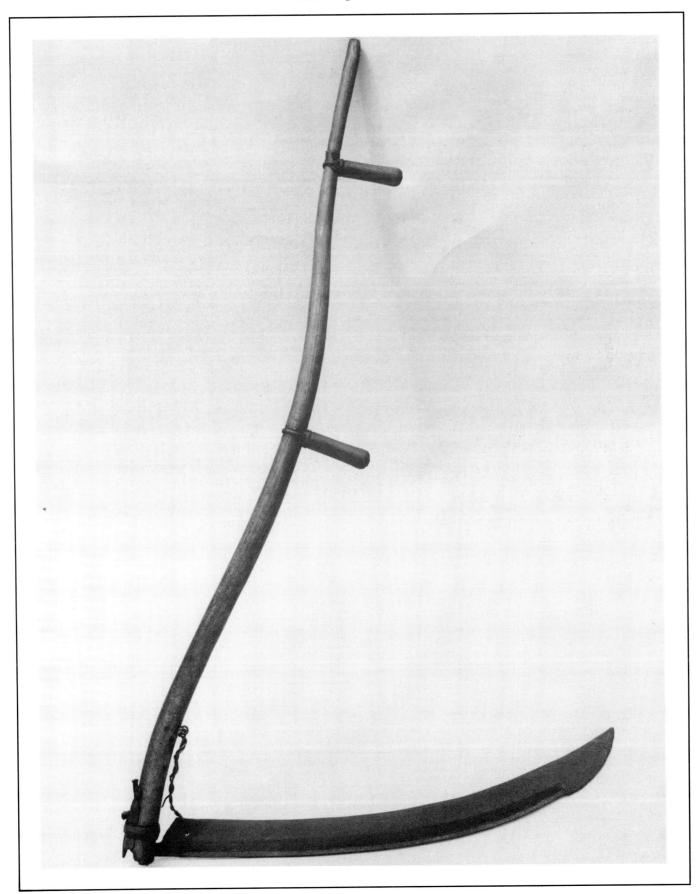

30 *A scythesman at work in Herefordshire in the 1930s. The looped hazel twig (bow) attached to the bottom of the sned was a device to help lay the cut corn evenly, and so make the work of the sheaf-gatherers easier.*

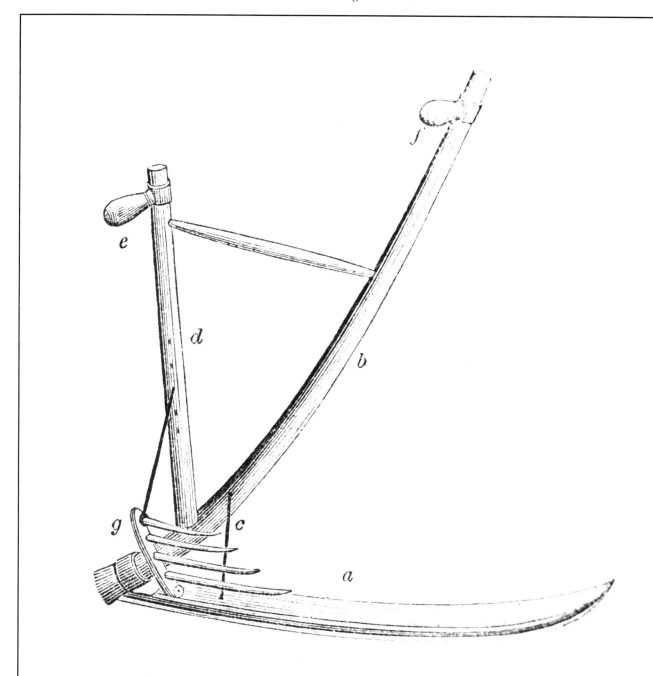

31 *A two-handled scythe, a type favoured more in
Scotland than England, showing another device for laying
the crop evenly. The wooden forks – often longer than on
this example – were known as a cradle, and common both
in England and Scotland. The illustration is from Henry
Stephens'* Book of the Farm.

32 *Sharpening the scythe. The sickle, with its toothed blade, did not need to be sharpened, but the smooth-bladed scythe required regular attention. The blade was sharpened with a stone or strickle. This tapering wooden blade was smeared with a mixture of grease and sand (carried in a small horn) which gave the required abrasion.*

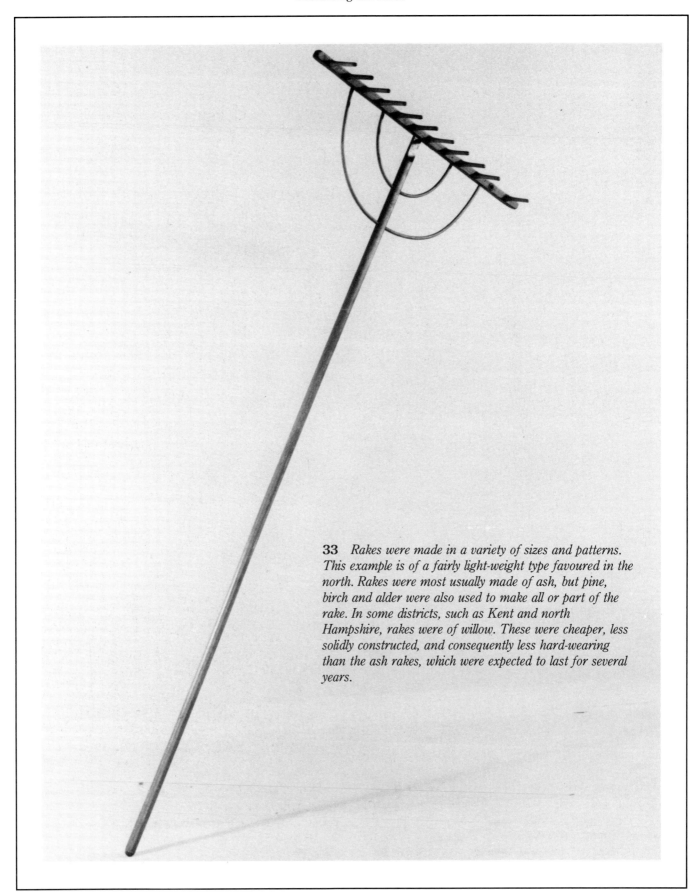

33 *Rakes were made in a variety of sizes and patterns. This example is of a fairly light-weight type favoured in the north. Rakes were most usually made of ash, but pine, birch and alder were also used to make all or part of the rake. In some districts, such as Kent and north Hampshire, rakes were of willow. These were cheaper, less solidly constructed, and consequently less hard-wearing than the ash rakes, which were expected to last for several years.*

34 *The rakemaker's yard. Although it was a simple tool, making a wooden rake was quite a complicated process. There were about fifty operations from the first cut of the wood to the fitting the final tine. In making the handle, which may appear to be but a length of wood, the wood* *first had to be cut to length and the bark peeled off, then made smooth, perfectly straight, and finally the ends shaped. The yard here is stocked with a variety of rakes and hay drags.*

35 *Making the tines. The wood was cut to a length of about six inches and the tines roughly shaped using billhooks and draw knives. To make them perfectly round a tine-former mounted on a horse was employed. The craftsman sat astride the horse, placed the rough-hewn tine into the top of the iron tube and struck it with a mallet to force it through the tube. The top edge of the tube shaved the tine smooth as it passed through.*

36 *Fixing the tines into the head of the rake. Rural rake-making survived into recent decades in some places, such as north Hampshire, where this photograph was taken in the 1950s.*

37 *Stooking always required careful placement. For a stook of six sheaves, the first two (in the centre) were placed perfectly upright leaning against each other. The outside pairs leaned inwards, their heads towards the centre. The man is shown using his knees as well as his hands to press the sheaf firmly into position.*

38 and 39 *Two types of stook are illustrated in Henry Stephens'* The Book of the Farm: *a standard stook of eight or ten sheaves and a hooded stook, a type most common in northern Britain and also in Kent. The extra sheaf laid across the top was intended to protect the ears of corn from the high rainfall experienced by most of these districts. The straw bands to tie the sheaves are also clearly illustrated.*

40 *Another method of protecting the corn against inclement weather was to build a large stook of twenty sheaves or more around a tripod – three poles tied together in wigwam fashion. Vents at the foot of the stook were left to allow air to flow through freely. It was a practice most common in the north and west, but it did reach other areas, especially in wet seasons. This example is from Oxfordshire.*

41 *A scene now vanished, but still common in 1948
when this was taken: the harvest landscape with well-made
stooks of wheat all across the field.*

42 *Pitching the corn for carting to the stack, in the early
1930s.*

43 *Building the corn stack, from a postcard posted in 1907.*

44   *Rick-building in Berkshire in the 1940s.*

45 *The method of building a circular stack illustrated in Henry Stephens'* The Book of the Farm, *showing the sheaves being placed in concentric circles, the outer sheaves presenting a wall of the butt ends. The man on the right is getting the sheaves firmly into place.*

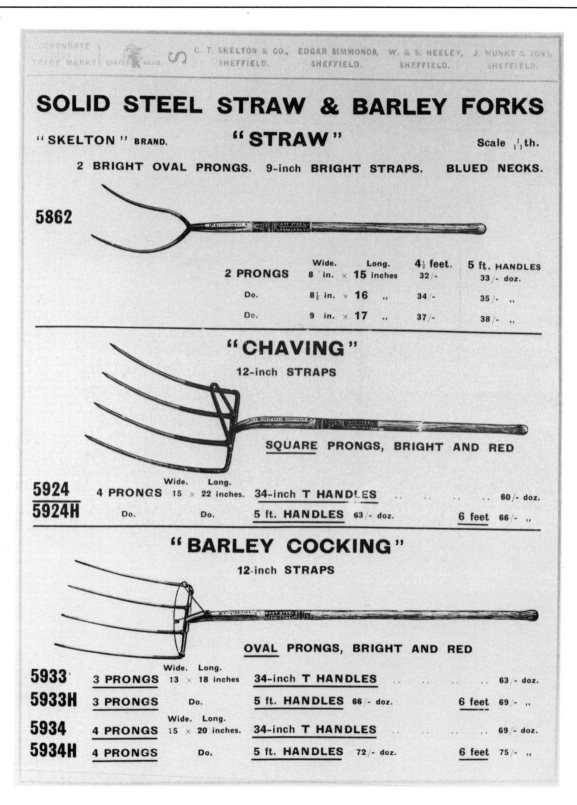

CORPORATE TRADE MARKS	QUAVE HEAD	S	C. T. SKELTON & CO., SHEFFIELD.	EDGAR SIMMONDS, SHEFFIELD.	W. & S. HEELEY, SHEFFIELD.	J. MUNKS & SONS, SHEFFIELD.

SOLID STEEL STRAW & BARLEY FORKS

"SKELTON" BRAND.   **"STRAW"**   Scale ¹⁄₁₁th.

2 BRIGHT OVAL PRONGS. 9-inch BRIGHT STRAPS. BLUED NECKS.

5862

	Wide.	Long.	4½ feet.	5 ft. HANDLES
2 PRONGS	8 in. × **15** inches	32/-	33/- doz.
Do.	8½ in. × **16** ,,	34/-	35/- ,,
Do.	9 in. × **17** ,,	37/-	38/- ,,

"CHAVING"
12-inch STRAPS

SQUARE PRONGS, BRIGHT AND RED

		Wide.	Long.					
5924	4 PRONGS	15 ×	22 inches.	34-inch T HANDLES	..	..	..	.. 60/- doz.
5924H	Do.		Do.	5 ft. HANDLES 63/- doz.			6 feet 66/- ,,	

"BARLEY COCKING"
12-inch STRAPS

OVAL PRONGS, BRIGHT AND RED

		Wide.	Long.					
5933	3 PRONGS	13 ×	18 inches	34-inch T HANDLES	..	..	..	.. 63/- doz.
5933H	3 PRONGS		Do.	5 ft. HANDLES 66/- doz.			6 feet 69/- ,,	
5934	4 PRONGS	Wide. 15 ×	Long. 20 inches.	34-inch T HANDLES	..	..	..	.. 69/- doz.
5934H	4 PRONGS		Do.	5 ft. HANDLES 72/- doz.			6 feet 75/- ,,	

46 *Forks for stacking and threshing, from the catalogue of C. T. Skelton & Co. The chaving and cocking forks had long, thin, widely-spaced tines, and were used, respectively, for gathering chavings (some of the refuse from threshing) and gathering short barley into heaps ready for pitching.*

47 *A stack built on staddle stones, Ebrington, Gloucestershire.*

PATENT CORN RICK STAND.

48 *Garrett's cast iron rick stand, from the firm's advertising of the mid-nineteenth century.*

49 *A stackyard on a farm managed by the Gloucestershire War Agricultural Committee in the 1940s. The wheat straw stacks are being thatched, with decorative finials added at the apex, while the tractor receives some running repairs.*

50 *During the Second World War Land Army girls took on many of the skilled jobs of the farm. Here one is combing the straw down on a rick being thatched in south Devon. The lengths of straw rope running across the thatch can clearly be seen.*

51   *The large needle thatchers used to sew the thatch on to a rick.*

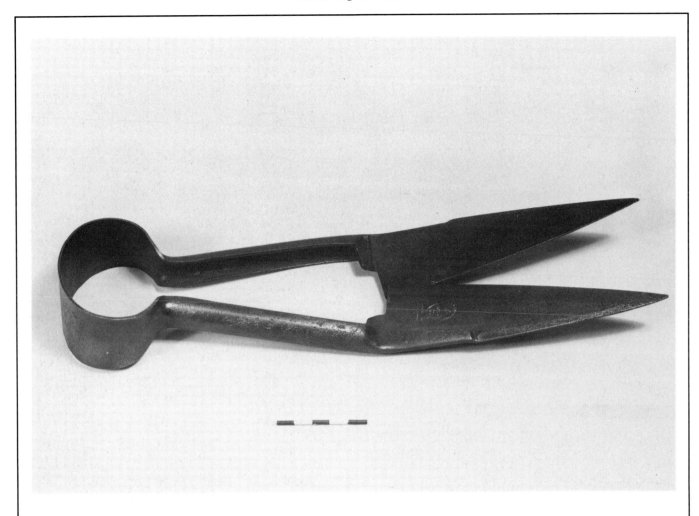

52 *Thatching shears were of similar pattern to sheep shears. As well as trimming the edges of the thatch, they were likely to be used for clipping hedges and cutting men's hair.*

53 *Making straw rope required two people: one walked backwards working the twister while the other fed straw into the rope.*

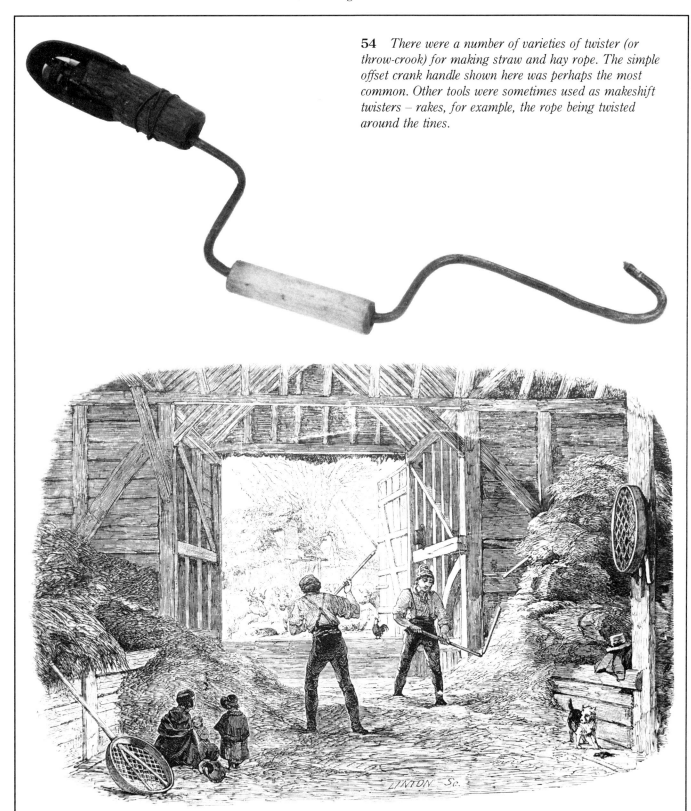

54 *There were a number of varieties of twister (or throw-crook) for making straw and hay rope. The simple offset crank handle shown here was perhaps the most common. Other tools were sometimes used as makeshift twisters – rakes, for example, the rope being twisted around the tines.*

55 *Inside the threshing barn, an illustration from the* Illustrated London News *of 1846. Boards were put across the doorway of the barn to keep out the livestock, shown taking a keen interest in proceedings inside. Riddles used for sifting out stray pieces of straw from the threshed grain are positioned at the side of the threshing floor.*

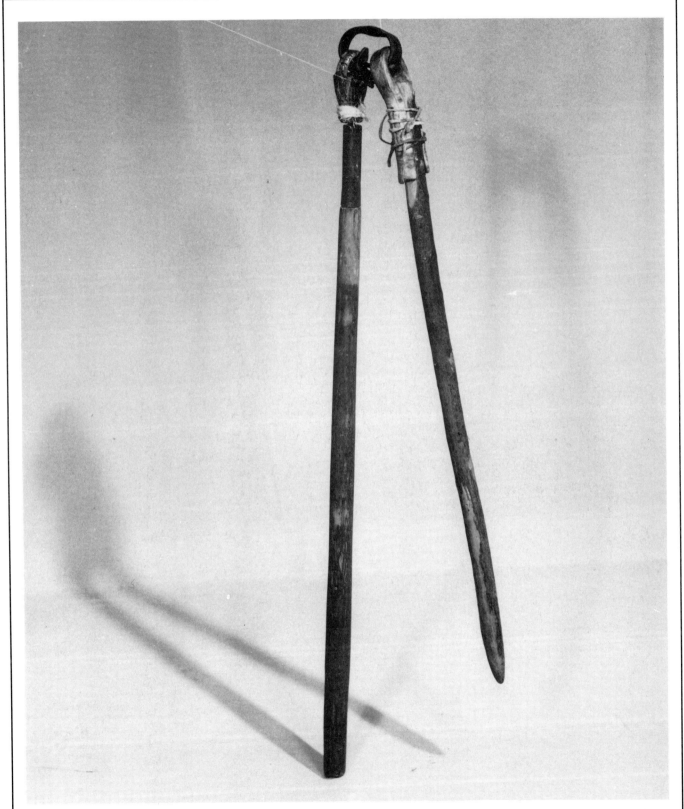

56 *A flail, showing the flexible joint between hand staff and beater. The hand staff, on the left, has a jointed head, and a cap of leather linked by a leather thong to a similar cap of leather or hide on the beater.*

57 *A winnowing basket thirty-nine-and-a-half inches long by twenty-eight inches wide. It has slats and handles of ash, and the body is of closely-woven willow basketwork, a type complicated and difficult to make.*

59 *The barley hummeller removed the sharp awns from the ears of threshed barley. There were two common types; a flat iron grid, with a handle about four feet tall, that was stamped up and down on the barley, and this roller type. The roller was the version preferred in East Anglia, and this example was used on a farm in Essex.*

58 *Grain shovels, used in the threshing barn, were of wood, being gentler on the grain. However, some, like this one, had an iron tip to the blade.*

60 *A prize ox, the Newbus Ox, looks on in anticipation
of the meal of roots being prepared with the guillotine-style
slicer in this print dated 1812.*

3

Farm Crafts in Animal Husbandry

The cowman had more routine to his life than the ordinary labourer working in the fields. He also had rather more security, for he, along with the shepherd and the carter (horseman), was likely to have been employed on a yearly engagement, whereas the ordinary labourers were employed by the day and might lose a day's pay if, for example, bad weather prevented work. The enhanced status of the workers with animals could be reflected in their pay. The Royal Commission on Labour in 1893, for example, recorded several districts where stockmen's earnings were one or two shillings a week higher than the day labourers'. But this was not always so: the commission's figures for Northumberland and Lancashire put stockmen's earnings at no more than the labourers'.

The cowman's routine varied according to whether his farm was mainly concerned with dairying or with raising beef. In dairying districts his was an early start to the day, cleaning the gutters in the cow byre, making the place presentable in readiness for the morning's milking. In winter, if the cows were kept indoors, there would be more of this work than in summer when the cows, if not milked in the field, were only brought indoors to be milked. The precise hour at which the cowman's work began was determined in many dairying areas by the time at which the morning milk train to London, Manchester or other large city left the local station. It would not be later than 6 am, often as early as 4 am.

A substantial part of the stockman's work was taken up with feeding the cattle and preparing the feed – slicing turnips, cutting chaff, breaking oilcake, preparing hay. The work grew as roots and other prepared feeds were given more commonly besides hay and pasture.

Roots had to be sliced up, and at the end of the eighteenth century this was often being done laboriously by hand, using a knife. Other simple hand tools and lever turnip slicers made the work somewhat easier. From the 1820s root choppers with revolving or oscillating knives powered by a simple hand crank improved productivity enormously. This type of machine could be powered by horse gear, steam engine or gas engine, but for many farmers into the mid-twentieth century the smaller hand-operated versions answered their needs.

Trussing Hay

Cutting hay from the stack required a fair amount of energy. The tools were simple: a hay knife and a hay pin. The knife had a large blade, shaped rather like a shield, and a double handle. Cutting the stack was effected by pressing hard down on the knife – no easy task through tightly compressed hay – for which good strong shoulders were needed. Progress in cutting a truss of hay was therefore steady rather than spectacular; repeated pressure on the knife pushing it an inch or two further each time, and gradually easing the hay away from the main body of the stack. The hay pin, meanwhile, had been rammed into the centre of the block being cut to hold it firm while it was carried down to the foot of the stack. The truss was tied using rope of hay or straw (always preferred over string made of textiles because the cattle or horses could eat it).

There were independent men who worked as hay trussers, travelling round the farms cutting down the stacks. Michael Henchard, the principal character of

Thomas Hardy's *The Mayor of Casterbridge*, was one of these. As well as cutting the stacks for use on the farm, these men were likely to be employed to truss up the stacks that had been sold to dealers in towns for feeding horses. These trusses had to be cut to the standard size – an oblong block about three feet three inches long by two feet two inches wide. Deviation from this size was likely to affect the price the farmer got from his sale. The hay trusser's work was confined mainly to autumn and winter. The independent workers would fill in the time in a variety of ways: harvest work, another craft or trade, or small-scale higgling or dealing. This was another feature that Thomas Hardy built into Henchard's character, who was able quite naturally to transfer to the role of corn dealer in Casterbridge.

Sheep: Feeding and Hurdles

The shepherd, like the stockman, had to spend time preparing feed for his sheep. Roots needed to be chopped and pulped, especially for the benefit of the old sheep that had lost some or all of their teeth. Fred Kitchen, in *Brother to the Ox*, recalls how the task of carrying skeps of sliced turnips across to the feeding troughs in the middle of a muddy sheep pen was far from the idyllic view of the shepherd's life.

Another part of the shepherd's work in all the lowland sheep farming districts was moving hurdles and feeding cages. Hurdles formed temporary fences to cordon off part of a field while sheep grazed there before being allowed to move on to fresh herbage. The farmers of southern England used them by the score on the downland pastures. They were also used to divide up the fields when sheep were folded upon fodder crops such as turnips and kale. They were particularly valuable at lambing time, when they could be set up to provide shelter. Portable feeding troughs and feeding cages were placed in the fields during winter and spring, filled with extra rations of hay or roots.

The hurdles and feeding cages themselves were supplied by craftsmen of the village and woodlands. There were two types of hurdle: the wattle hurdle and the gate hurdle. The wattle hurdle was made of seven-year-old hazel coppice. The hurdle maker harvested his hazel during the winter in preparation for the summer months spent in making the hurdles. This was usually done out in the open in the woodlands, with at most a rudimentary shelter for inclement weather. The hurdle maker's tools were few, no more than a few knives for cleaving and trimming the hazel, and the 'mould' – a log about seven feet long, with ten holes drilled in line, into which the the uprights of the hurdle were inserted while the horizontal strips were woven around them.

The bottom ten inches of the hurdle were woven with rods in the round, using the long, thin ones no more than three-quarters of an inch thick. These were known as spur rods, and gave a strong, firm base to the hurdle. Most of the hurdle was then woven with cleft rods, some more uncleft ones making up the top two or three inches.

About a dozen hurdles a day was the wattle maker's output. The hurdles were not intended to be long-lasting – about three years was common. Work, therefore, promised to be quite steady, until, that is, the decline in arable sheep farming set in during the twentieth century.

Gate hurdles, similarly, were not made for a long life, although some were quite substantial in appearance. Willow was the favoured wood, being light – an important consideration for hurdles that had to be carried about the fields – and it could be split easily by cleaving or riving. Ash was the second choice and birch was sometimes used.

The gate hurdle maker usually worked in his workshop in the village. Hurdles were all of a similar pattern, although differing slightly from district to district. Five, six or seven horizontal rails were fitted between two sturdy uprights (known as heads). A third upright strengthened the centre, and a pair of diagonal braces was added. Hurdles varied slightly in size from place to place. In Hampshire the standard was six feet long by three feet six inches high. Gloucestershire hurdles were longer: seven feet six inches.

Sheep-shearing

Shearing time was one of the peaks in the year for the sheep farmer. It was common to wash the sheep before they were sheared. In many places this took on something of a communal nature, where there was only one stream in a parish suitable for the washing. The farmers all brought their sheep down to the stream where they were penned until they were allowed into the water one at a time.

Shearing could also have its communal aspects, as neighbours got together and clipped each other's flocks in turn. This was the tradition, for example, amongst the hill farmers of Wales, and in parts of England where flocks were of modest size. Many farmers, especially those with large flocks, preferred to employ the specialist shearers who toured the countryside. Rider Haggard describes the arrival at his farm in Norfolk of the 'gang of shearers, four in number, who travel with a pony and cart from farm to farm, clipping the sheep at a charge that averages about threepence a fleece'.

Sheep shearers worked one man to a sheep. With the sheep sitting on its tail, the shearer started clipping at the belly, working round towards the back. The animal was turned on to its side for the back to be clipped; the sheep's legs also were tied together with thin cord to prevent its trying to scramble up. Shearing was performed entirely with simple hand shears until the end of the nineteenth century. Mechanical clippers were introduced from Australia in the early 1890s, but seem to have found relatively little favour for many years. These early clippers were powered by a hand-crank mechanism. Only after power from gas engines (and later electricity) was applied did mechanical clippers begin to make headway. Hand shears remained common into the 1940s.

After the shearers had done their work, the fleeces were weighed and packed (compressed as tightly as possible) to be sent away to the wool dealer.

61 *Sheep folded in a sheltered area formed by a fence of wattle hurdles. The wooden feeding cages provide additional rations.*

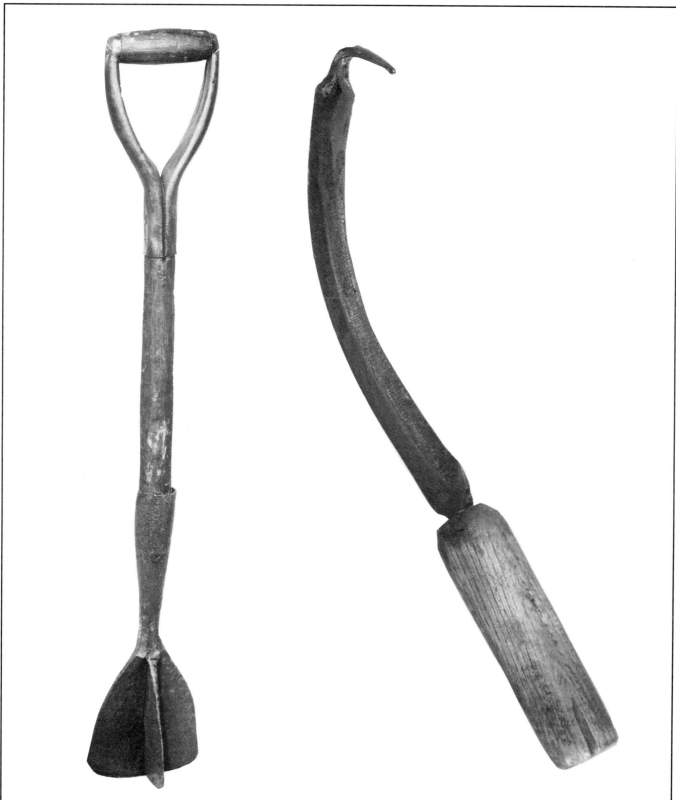

62 *A turnip chopper, a simple tool that would cut a turnip into four pieces. These choppers were still being made by the leading makers of edge tools in the 1920s.*

63 *A turnip knife, this example coming from a farm in Lincolnshire.*

64 *A spale feeding basket, a type common in much of north and midland England, including Worcestershire, Shropshire, Derbyshire, Yorkshire, the Lake District and Lancashire, from where this one comes. It is of oak spale with a willow rim. The oak is quartered, then split further into finer bands, using a sharp knife (called a lath axe) and a mallet. These spelks are trimmed with a spokeshave. They have to be kept moist in order to be pliant for weaving.*

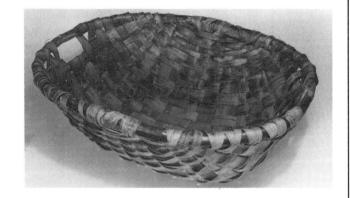

65 *In Wales this oval form of basket was common, both on the farm for feeding cattle, and for shopping. The frame is made of hazel strips around which willow is plaited.*

66 *Sutton & Sons, the seedsmen, took this photograph to show the prolific yield of their Golden Tankard mangels. It also shows the uphill task faced by the two men with the root slicer. They are using one of the machines with rotating knives operated by cranking the flywheel at the side. It could cut rather more roots than the simple choppers and knives.*

67 *A chaff-cutter illustrated in the* Microcosm *by W. H. Pyne (1803). The man cuts with the guillotine knife while pushing the chaff forward in the box with the rake in his left hand.*

68 *Cutting hay from the stack. The hay pin is plunged into the stack by the man's right leg.*

69 *A Brades 'Crown' hay knife, a type that the firm included in catalogues into the mid-1950s.*

70 *Taking hurdles out to the field. Although hurdles were reasonably light in weight, intended to be carried in this way, it was, even so, hard physical work, putting strain on the back and knees.*

71 *A wattle hurdle maker, photographed in 1890, at work out in the open, his stock of hazel rods and trimmings behind him.*

72 *The work of weaving the hurdle is shown in closer detail here. The bottom section of the hurdle, woven with round rods, is complete, and the maker is weaving the hazel strips that make up the central part, around the uprights.*

73 *Making a gate hurdle. A willow pole is cleft into two or three strips to make the horizontal rails. The pole is clamped into a brake and split with a tool known as a froe or dill axe, which is shaped similarly to an axe, but with the cutting edge on the top of the blade instead of the side. After cleaving, the ends of the rails are shaved flat where they will fit into the upright heads, and the bark is planed off.*

74 *Assembling the hurdle. The pieces are laid out on a frame to be put together. The head in the foreground shows the mortices where the rails are to be fitted.*

75 *Sheep feeding cages were usually made of hazel coppice split into strips and bent into shape. Here some of the final nails are being added to an almost complete cage.*

Figs. 2 and 3—exhibit the two ends of each Hurdle when made of Wood, with the socket end for the reception of the Hook attached; these may be made about 18 feet long.

Fig. 4—represents the mode in which the Hurdles may be constructed with wire bars.

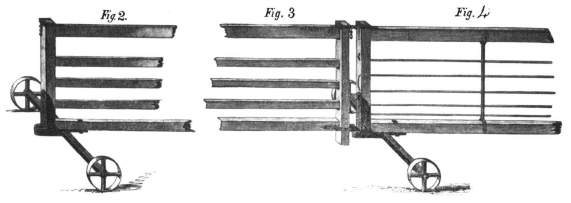

Fig. 2.   *Fig. 3*   *Fig. 4.*

Fig. 5—is the Rack and Trough, with iron ends and wheels.

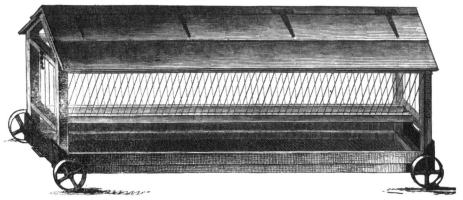

Fig. 6.—Hook for the light end of the Hurdle.

Fig. 6.

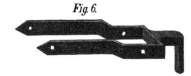

Fig. 7.—Wrought-Iron Ferrules for the top of the uprights, with chain and hook to attach each to the other.

Fig. 7.

Fig. 8.—Wrought-Iron Upright, used to strengthen the Hurdles when constructed of Wire.

Fig. 8.

76 *Agricultural implement manufacturers soon entered the market for feeding equipment. Ransomes of Ipswich* *were advertising these iron feeding cages on wheels in the 1840s.*

77 *About a fortnight before they were sheared sheep were washed in a running stream to get the dirt and grease out of the fleece. The men on the banks guided the sheep with their poles while one man got into the water and gave each sheep a good rubbing down to make sure it was clean. By the time this photograph was taken in the 1930s the practice was well into a decline. A clean fleece fetched a slightly higher price, but that was offset by the fact that it weighed less. Most farmers were concluding by this time that the lost weight outweighed the gain in price.*

79 *A gang of sheep shearers in Dorset working in the open, with large cloths spread out as ground sheets and hurdles set up as protection from the wind.*

78 *The shearer, wrote Rider Haggard in* The Farmer's Year, *seizes the ewe, 'and with an adroit and practised movement causes her to sit upon her tail . . . The operator begins his task in the region of the belly, working gradually round towards the back until it is necessary to turn the animal on to her side, when he ties the fore and hind leg together with a thin cord.'*

80 *Catalogues for Thomas Bigg's sheep dips in the late nineteenth century included this illustration of dipping in progress. 'Two men seize the sheep one by one and plunge them legs upwards into a v-shaped tub half full of unpleasant-looking fluid', was Rider Haggard's description of the work.*

81 *Statutory regulation of dipping, which decreed that each sheep must be immersed for one minute, encouraged the change to baths sunk into the ground. The police constable was in attendance to see that the law's required time was observed. This photograph was taken in Yorkshire in the mid-1930s.*

4

Maintaining the Farm

Hedging and Ditching

The winter months of the year were the time when the hedges were cut back and the ditches cleaned. As with so many other tasks, these were often entrusted to independent workers who went from farm to farm, carrying their own tools. Joseph Arch, who became leader of the National Agricultural Labourers' Union in the 1870s, worked as a hedger throughout Warwickshire, Gloucestershire, Herefordshire and the Welsh border counties.

Over the greater part of England hedges were usually maintained in a relatively simple manner by cutting back and pruning with shears and billhooks. The trees of the hedgerow (most commonly elm, oak and ash) were maintained, and gaps in the hedge filled with new planting or by training the existing growth.

The midland counties were the home of the tradition of hedge laying, whereby the main stems and branches were trained laterally at an angle of about sixty degrees. The aim was to encourage the branches to tiller, especially at the bottom, so forming a neat and effective barrier to livestock. During the twentieth century rural training, craft revival schemes and competitions have encouraged the practice of hedge-laying to spread more widely.

There were many varieties of practice in laying hedges, from region to region and between individuals. The basic principles were common to all. The first task was to cut away all the undergrowth and the loose, small pieces along the sides. Unwanted plants such as bramble and elder were cut out of the body of the hedge. Now the real work of laying the hedge could begin. The main stems of the hedging trees were cut with the billhook, about one foot above the ground, taking care not to cut right through, for

the plants must continue living. These cut stems (pleaches) were then laid, bending them back all in the same direction (usually to the left). At intervals of approximately two feet there was a vertical post to help bind the pleaches together. Where there was a conveniently situated sapling, that would be cut, otherwise stakes would be driven in. A binding of rods (usually hazel or elm) woven together and around the upright posts might be added along the top to add rigidity to the hedge.

Dry-stone Walling

Dry-stone walls are a prominent feature of those parts of northern England, Wales and the Cotswolds where large stones are plentiful on the surface of the ground. Building the wall requires particular skills in distributing the weight of the stones so that they bind together without mortar to produce a wall strong enough to withstand sheep and high winds. These are not the skills of the mason or builder. Henry Stephens in *The Book of the Farm* complained, 'We suspect that many dry stone dykes are built by ordinary masons, who, being accustomed to the use of lime mortar, are not acquainted with the proper method of bedding down stones in a dry dyke as firmly as they should be, and are therefore unfitted to build such a dyke. A builder of dry stone dykes should be trained to the business, and with skill will build substantial dykes at a moderate cost which will stand erect for many years.'

The walls differ in appearance from region to region, but the principles for building them are universal, relying on the careful selection and placement of stones of different sizes. In building a new wall, its line was first marked out and a shallow trench cut along that line. Wooden frames were placed at intervals to mark out the shape of the wall, with strings running between the frames.

The largest, more regular stones formed the foundation for the wall. The rest of the wall was built up from that base. Tie stones were inserted at intervals, these being large stones laid lengthways across the line of the wall, helping to bind the others together. In the narrower upper parts of the walls the larger tie stones would be allowed to protrude to either side (this gave them the name 'throughs').

Henry Stephens again noted the importance of carefully placing the stones as the wall was built up: 'Great art is required in laying the small stones, and it is this in dyke-building which detects the good from the bad dyker. In good dry-building, the stones are laid with a slight inclination downwards, from the centre of the dyke towards each face . . . and to support the inclination, small stones are wedged firmly under them in the heart of the dyke; whereas stones laid flat admit of no wedging to firm them, and receive thorough-bands.' The top of the wall was levelled and finished with a row of coping stones ('combers' in the Cotswolds).

The stone of the Cotswolds is reasonably easy to cleave, and here the waller might shape his stones using a heavy hammer with two cutting edges. In general, however, walls were built with the irregularly shaped material to hand, each stone being chosen to fill a gap.

82 *A ditcher at work, from a photograph dated 1910.*

84 *Gathering the hedge trimmings on to a sledge in Westmorland in the early 1940s.*

83 *The hedger, billhook in hand. He is wearing the thick leather gloves that afforded protection from hawthorn, bramble, sharp twigs and stones. The photograph was taken in the 1890s.*

85 *A neatly laid hedge in Warwickshire, with the hedger doing some final trimming at the base.*

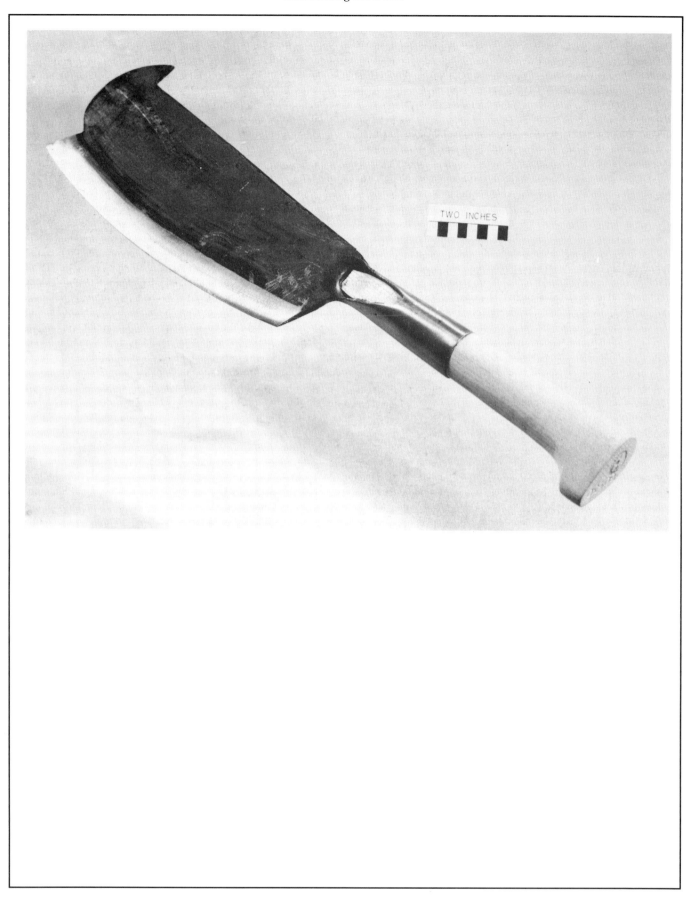

86 and 87 *Two billhooks made by the firm of A. Morris & Sons, Dunsford, Devon: 86 was sold as a Hertfordshire socket bill hook, having its steel blade fitted into the ash handle with a socket; 87 was described as a Devon hook.*

88 *Setting the coping stones on a dry stone wall at Parwick, Derbyshire, in the late 1930s.*

89 *Dry stone walling as illustrated in Henry Stephens'* Book of the Farm, *showing the wooden frames and marker strings, with the stages of construction within that frame.*

5

Agricultural-Servicing Crafts

Henry Stephens wrote: 'Iron, wood and leather are the materials of which the implements of the farm are constructed; and the implements, being in constant use, are in a continuous state of decay, and require daily repair. To effect constant repairs it is impracticable to send implements to the nearest town or village. Hence the necessity of having a smithy and joiner's shop on the farm.'

This represents an ideal few farmers were likely to attain. Farmers did not lavish quite the daily care and attention in keeping their tools and implements repaired that Stephens suggested they should. Few farms were able to equip themselves with a workshop for smith or carpenter, although it was not at all uncommon for such craftsmen to go out to the farms to do work. Even so, Stephens' remarks demonstrate how important to farming such servicing crafts were, how dependent farmers were upon these crafts, in particular the blacksmith, wheelwright and carpenter.

Blacksmith

It is apparent from surviving account books that the blacksmith was a most important member of the farming economy. In the middle of the eighteenth century, for example, the blacksmith at the village of Bucklebury in Berkshire was John Hedges. His family had been smiths there for 150 years, and succeeding generations were to carry on the business for another 150. Almost all of John Hedges' customers were farmers. They brought their horses to be shod and their ploughs to be repaired. Ploughshares and coulters were brought in regularly to be reshaped, sharpened and repointed. These, the cutting blades of the plough, were costly to buy new in 1750. A share cost at least nine shillings, an expense the farmer was happy to put off as long as possible by getting the smith to sharpen the tool, at ten pence a time.

Sharpening shares and shoeing horses accounted for about half of Hedges' work. Work on wagons, mainly tyring the wheels, also occupied much of his time. The farmers also brought him all their hand tools to be repaired. There were spades, rakes, shovels, forks, billhooks, mattocks, turnip knives, weeding tools (hoes, thistle spuds, dock pullers) scythes and reaphooks. All these had blades and prongs to be sharpened, put right and occasionally replaced. The cost of iron was such that even the bolts and pins holding the tools together were sometimes considered worth repairing rather than replacing.

What was true of John Hedges was true also of the Dale family, smiths at Little Warford in Cheshire. Their trade at this time came from the sale and repair of pitchforks, scythes, rings for pigs and bulls, spades, knives, horse combs and all the other tools and implements of farming. The experience was repeated countless times, for the rural blacksmiths scattered thoughout the villages of England were primarily working for farmers. Many were shoeing smiths only. A hundred years later, in the mid-nineteenth century, smiths were still working mainly for farmers. The jobs they were doing were not all the same. They were less likely to spend time sharpening ploughshares now that cast-iron self-sharpening shares were in common use. But, with many more iron parts on ploughs than on the old wooden-beamed types, there was plenty of other repair work. In addition, there were all the other large implements introduced during the nineteenth century that had iron fittings to be repaired. Chain harrows would need new tines.

Seed drills had iron coulters and gearing to the seed feeding mechanisms were likely to break. There were tines to fix on heavy cultivators, cutting blades on mowing machines and mechanical parts on threshing machines. Alongside that there remained work on the scythes, billhooks, hoes and other hand tools and, of course, the shoeing of horses, which continued to be the bread and butter of many a smith's work.

All of this meant that the blacksmith's work was growing during the mid-nineteenth century, enough to keep more smiths at work. The census returns for Berkshire, for example, record two hundred more blacksmiths in 1871 than in 1851. This proved to be something of a heyday for the country blacksmith. Declining prosperity in farming in the later decades of the nineteenth century made farmers more sparing in their demand for repair. New implements of the late nineteenth and twentieth centuries increasingly required the services rather of the mechanic than the smith to keep them in working order. The numbers of horses in farming began to fall. So, too, did the number of smiths, as they retired or turned their businesses into garages or agricultural-servicing agencies.

Wheelwright

'In the timber room of the horse buildings, Robson, the labouring carpenter, is engaged in mending the wheels of an old pony-cart. Now, to make a wheel, or even to set some spokes in it, is a thing that looks easy, but, as a matter of fact, it demands much skill and practice. This particular wheelwright is only a hedge carpenter, without even a shop of his own, but he has the reputation of being able to "set" a wheel better than anyone about here, and certainly his work is always very sound and good.' Rider Haggard in the 1890s was describing a tradition that lingered on from an earlier age. The jobbing, hedgerow carpenter had in the eighteenth century made some of the wooden-beamed ploughs of that period, and repaired many more. In addition he repaired the gates, pig troughs, mangers and other woodwork of the farm.

By contrast, the wheelwright's was a more superior business, being based usually in a workshop and having as its main trade the repair of wagons, the making and repair of wheels. For many small wheelwrights in the villages there was, however, little difference between them and the humble hedge carpenter. The wheelwright, too, did general carpentry of all types. Repairs provided the bulk of his trade. With exceptions, it was not often that he made a complete new wagon. That type of work tended to go to the larger firms of wheelwrights in the market town and, as the nineteenth century wore on, to the yet larger firms of factory wagon builders, such as the Bristol Wagon Works.

Spring was often one of the busy times of the year for the village wheelwright. It was then that farmers preparing for the spring cultivations and sowing pulled their carts and wagons out of the sheds only to discover defects that should have been attended to in the autumn. So, the wagons were brought down, rattling and creaking, to the wheelwright's shop with urgent requests for new floorboards, repairs to the tailgate, some attention to the wheels or a new coat of paint and varnish. Along, too, came the seed drills with the broken shafts, even the small tools such as mattocks and forks in need of a new handle. For a few weeks the wheelwright had to work furiously to get all the jobs done on time.

As Rider Haggard noticed, the making and repairing of wheels was one of the most skilled of tasks. There were only three main components to a wheel: the central hub or nave, spokes, and the rim made up of a number of sections called felloes. In the shaping and setting together of these lay a great deal of skill. The hub was made of elm, a solid block shaped to its final diameter of twelve to eighteen inches on a specially designed type of lathe. Spokes were made from pieces of oak, cleft along the grain to preserve the strength of the wood. These pieces were roughly shaped, then finely pared with draw knife and spokeshave. They were driven into the mortices cut into the hub using a fourteen-pound sledgehammer. The felloes were of ash, about thirty inches long and three-and-a-half inches square, smoothed to the correct shape with adze and jack plane. Each felloe covered two spokes – so wheels always had an even number of spokes.

The completed wheel was shod with an iron tyre or with strakes. Strakes, individual strips of iron nailed on to the wheel, were fixed cold, but hoop tyres, which became steadily more common from the late eighteenth century, were put on hot. This was, strictly, blacksmith's work, and tyring a wheel required co-operation between smith and wheelwright. So close might this be that, as George Ewart Evans discovered in many villages in Suffolk, the wheel-wright's shop and blacksmith's forge would often be almost next door to each other, and there were numbers of combined businesses.

The middle decades of the nineteenth century represented a high-water mark for many of the crafts and techniques described in this book. The numbers of black-smiths and wheelwrights, which had been rising in many parts of the country after 1851, were declining again by the 1890s. The effects of industrialisation and standardisation were being felt. Several of George Ewart Evans's respondents in Suffolk would say that 'blacksmith-made' was always the highest accolade that could be accorded a tool but, as the years of the twentieth century advanced, relatively few were, in fact, using such tools, opting instead for the factory-made product.

Mechanisation and the depressed state of much of agriculture in the late nineteenth century and between the world wars both had their impact. If the flail was already a memory by the 1890s, it was being joined by the sickle, the bagging hook and the scythe during the following decades, while such techniques as rick thatching began to decline. Not all have completely disappeared: hedge-laying and dry-stone walling have retained a place in the countryside.

90 and 91 (over)   *Two views of smith's shops, inside and out, taken in the early years of the twentieth century, showing the accumulation of farm implements awaiting attention, and work around the forge.*

93 *A wheelwright in Herefordshire in about 1933 doing some repairs to a wheel.*

92 *One way by which smiths might meet the demands of farmers for work at the farm was a portable forge. This Edwardian photograph shows the assistant tending the forge while the smith is working on a tool behind him.*

94 and 95 *A hoop tyre was fitted to a wheel by being first heated so that the metal would expand sufficiently for the tyre to slip round the rim of the wheel. Then it was cooled rapidly causing it to shrink and fit tightly round the wheel. This was a job usually done outside. Many smiths had a special tiring platform in their yard, others simply used the open yard, even occasionally the road outside, in which to work. All hands that smith and wheelwright could muster between them were employed to lower the hot tyre, gripped firmly with tongs, round the wheel and then to dowse it with water to cool it.*

Bibliography

R. N. BACON, *Report of the Agriculture of Norfolk*, 1844

E. J. T. COLLINS, *Sickle to Combine*, 1969

R. W. DICKSON, *The Farmer's Companion*, 1813

J. DONALDSON, *British Agriculture*, 1860

GEORGE EWART EVANS, *Ask the Fellows who Cut the Hay*, 1956

GEORGE EWART EVANS, *The Farm and the Village*, 1969

W. FREAM, *The Elements of Agriculture*, 1893 edition

H. RIDER HAGGARD, *A Farmer's Year*, 1899

T. HENNELL, *Change on the Farm*, 1934

J. G. JENKINS, *Traditional Country Craftsmen*, 1965

CHARLES KIGHTLY, *Country Voices*, 1984

PRIMROSE MCCONNELL, *The Complete Farmer*, 1911

WILLIAM MARSHALL, *Rural Economy of the West of England*, 1796

HENRY STEPHENS, *The Book of the Farm*, 2nd edition, 1852; 3rd edition, 1877

C. HENRY WARREN, *Corn Country*, 1940

JOHN M. WILSON, *Rural Cyclopedia*, 4 vols, 1847–52

ARTHUR YOUNG, *General View of the Agriculture of Lincolnshire*, 1813

ARTHUR YOUNG, *General View of the Agriculture of Suffolk*, 1813

Index